THE COSMOS

A Complete Book on the Universe, Existence, and Meaning

Title: *The Cosmos: A Complete Book on the Universe, Existence, and Meaning*

Author: Joseph Farnandes

PREFACE

This book took a long time to write. It almost had to.

The subject itself resists speed. The universe doesn't give up its meaning to rushed attention, or to people who think the right answer should arrive quickly if you just look hard enough. It stretches across distances that break intuition and across spans of time that make human urgency feel oddly out of place. To engage it honestly requires patience, restraint, and a certain tolerance for questions that don't resolve neatly.

I didn't begin this book to prove a point. I wasn't trying to defend a doctrine, settle an argument, or reconcile every tension between science, philosophy, and faith. I began it for a simpler reason, though not an easier one: because looking up at the night sky has always felt less like a problem to be solved and more like an invitation—one that keeps returning, no matter how much we think we've learned.

We live in an era of astonishing knowledge. Humanity knows more than ever before about the origin of the universe, the structure of matter, the evolution of life, and the long future of cosmic time. We can trace the universe back toward its earliest moments and forward into eras far beyond the lifespan of stars. And yet, alongside this expanding knowledge, a quiet confusion remains.

Some feel that science has drained the universe of meaning. Others feel that faith must shrink or retreat in the face of evidence. Still others find themselves caught somewhere in between—drawn by wonder, unsettled by doubt, unsure where they are allowed to stand without betraying either honesty or hope.

This book is written for those readers.

It is for the curious who resist shallow certainty, for the thoughtful who refuse to choose between intellectual seriousness and humility, and for those who sense that explanation alone does not exhaust reality. It assumes that science deserves respect, that faith deserves careful reflection, and that philosophy deserves patience. It also assumes that none of these, taken in isolation, can fully account for what it feels like to be human in this universe.

Throughout these pages, I've tried to let each domain speak in its own voice. Science describes the universe as it finds it, without being forced into metaphysical conclusions. Philosophy asks questions that science does not answer, without dismissing scientific insight. Faith reflects on meaning and purpose, without claiming authority over empirical facts. Each has its limits. Each also has something essential to offer.

This approach will disappoint anyone looking for easy answers.

It won't offer comfort through oversimplification, or certainty through denial of complexity. What it offers instead is quieter, and I hope more durable: a way of standing within the universe that is intellectually honest and existentially serious at the same time.

If this book succeeds, it won't close conversations. It will deepen them.

And if, after reading, you find yourself looking at the sky with a little more attention, asking better questions, and feeling a more grounded sense of responsibility toward life and meaning, then it will have done what it was meant to do.

INTRODUCTION

Before there were theories, there was wonder.

Long before equations described the expansion of space or experiments revealed the structure of atoms, human beings stood beneath the sky and asked questions with no immediate answers. What is this vastness above us. Where did it come from. Why does it exist at all.

And what, if anything, does our own existence mean within it.

Those questions never disappeared. They just learned new languages.

Modern science has reshaped our understanding of the universe in ways earlier generations could hardly have imagined. We now know that the universe has a history, that it expanded from a hot and dense beginning, that stars forge the elements life depends on, and that time itself has a direction. We can measure distances across billions of light years and detect light that has traveled nearly since the beginning of everything we can observe.

And still, the mystery hasn't faded. If anything, it has grown.

The universe turns out to be far larger, older, and stranger than we once believed. Most of it is invisible. Its laws are precise, yet their deeper origin remains unexplained. Its future stretches beyond every familiar structure we know. And within this immense, restless reality, life and consciousness have appeared—at least once—giving the universe the strange ability to reflect on itself.

This book is about that story.

It is about the cosmos as science reveals it, but also about what it feels like to live inside such a cosmos. It asks how order emerges, how complexity becomes possible, how life and mind arise from matter, and how meaning can survive in a universe that offers no obvious guarantees. These questions aren't treated as separate, because they aren't separate in lived experience.

How we understand the universe shapes how we understand ourselves.

When Earth is no longer the center, humility becomes unavoidable. When time stretches across billions of years, urgency changes its character. When life appears fragile—and possibly rare—responsibility deepens. And when consciousness emerges from matter, questions of value can no longer be brushed aside as mere illusions.

This introduction marks a threshold.

Beyond it lies a shift from everyday intuition to a cosmic perspective. The chapters that follow move from the birth of the universe to the laws that govern it, from stars and galaxies to life and consciousness, and finally toward questions of faith, purpose, and meaning. Each chapter approaches its subject differently, because the universe itself does not repeat the same pattern at every scale. The structure of the book follows that diversity, letting form respond to insight rather than habit.

You don't need specialized training to read this book.

What you need is curiosity, patience, and a willingness to sit with complexity without demanding immediate resolution. Some answers will come into focus. Others will remain open. Both are part of honest inquiry.

The universe is not a puzzle meant to be solved once and set aside.

It is a reality to be inhabited, thoughtfully and responsibly.

This book invites you to step into that reality with open eyes, intact questions, and a sense of wonder that hasn't been rushed out of existence.

PROLOGUE: Standing at the Edge of Wonder

Look up.

Seriously. Just for a second. Before formulas, before telescopes, before anyone figured out how to put names on distant lights—there was this one simple thing humans have always done. We lifted our heads and stared into the dark.

People have done this for tens of thousands of years. On frozen plains. In deserts. On mountains. On rooftops in cities that didn't exist yet. Different languages, different myths, same feeling every time. That quiet jolt in the chest. The universe is enormous. We are tiny. And somehow... we're here noticing it.

That's where this book starts. Right there.

Every culture that's ever existed has asked some version of the same questions. What is all this? Where did it come from? Why does it exist at all? And maybe the strangest one—why are *we* here inside it? Those questions didn't come from textbooks. They came from being human. They show up on their own, the way hunger or thirst does.

Modern cosmology has answered a lot. More than most people realize. But it's also made things weirder.

We now know the universe has a history. It wasn't always like this. It used to be hotter. Denser. Smaller—so small it's hard to wrap your mind around. Time itself seems to have a starting point. Space stretches. Galaxies rush away from each other. Forces we can't see shape everything we can. And somehow, out of all that chaos and motion, stars formed. Then planets. Then life. Then minds capable of wondering how any of it happened.

That last part should stop you for a moment.

You're not outside the universe looking in. You're part of it. You're something it's doing. The atoms in your body were forged in stars that lived and died long before Earth existed. The calcium in your bones, the iron in your blood, the oxygen in your lungs—all of it came from ancient cosmic furnaces. Matter figured out how to organize itself. Chemistry stumbled into thought. And now the universe is aware of itself, at least a little, through human consciousness.

That might sound poetic. It is. But it's also just... true.

This journey is going to move across absurdly large scales. We'll dip into the smallest things we can describe, where particles blink in and out of existence, and then pull back to the largest structures imaginable—galaxies strung across space like filaments, spanning billions of light-years. We'll talk about the universe when it was less than a second old. We'll also talk about futures so distant that even stars eventually burn out and disappear. There will be equations. There will be evidence. There will be experiments. But there will also be questions no instrument can fully answer.

Because facts alone don't satisfy us. They never have.

Science is incredibly good at explaining *how* things behave. It finds patterns. Laws. Histories. It does this with stunning precision. But when we start asking why there's something instead of nothing, or why the laws are the way they are, or why consciousness exists at all—things get less tidy. Philosophy has wrestled with that terrain for thousands of years.

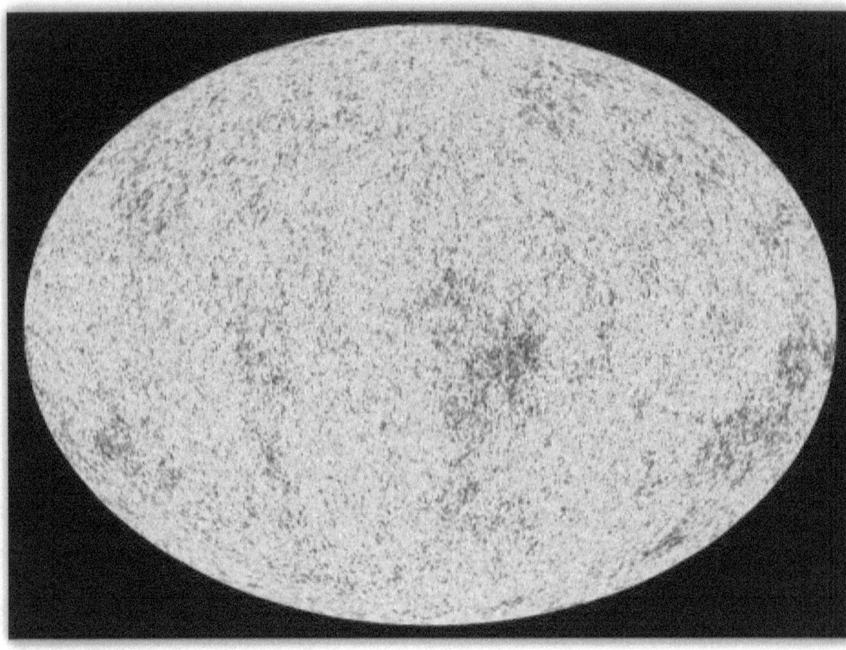

Theology has approached it through faith and reflection. Humans have always wanted more than explanation. We want meaning.

This book isn't asking you to pick sides. Not between science and faith. Not between reason and wonder. Not between curiosity and humility. Reality is big enough to require all of them. Where science speaks clearly, we'll listen. Where it goes quiet, we won't rush to plug the silence with easy answers. Mystery, when handled honestly, isn't a failure. It's often an invitation.

We're living at a strange and remarkable moment. For the first time, one species on one small planet has managed to piece together a reliable story of the entire cosmos—from its earliest moments to its possible ends. We can measure the universe's age. We can map the faint afterglow of its birth. We can find planets orbiting stars we'll never visit. And somehow, every discovery just opens more questions. The universe doesn't shrink as we understand it. It expands.

There's a kind of responsibility hidden in that understanding too. Seeing the universe clearly changes how you see yourself. Small arguments feel smaller when you think in cosmic time. Arrogance doesn't survive well in deep space. And yet—human life becomes more precious, not less. In a universe that seems indifferent, the emergence of consciousness, compassion, and moral choice is astonishing.

This book is an invitation. Not to certainty, but to curiosity. Not too easy answers, but to honest questions. You don't need a background in physics or philosophy. You just need the willingness to wonder—and to sit with what you don't know.

We'll explore the birth of the universe, the structure of galaxies, the strange rules that govern reality, and the emergence of life and mind. We'll confront the possibility that the universe is balanced in ways that allow us to exist, and ask what—if anything—that means. We'll talk about faith not as an escape from reason, but as one of the ways humans respond to depth, order, and mystery.

In the end, this isn't really a book about distant stars. It's about you. About what it means to be a thinking, feeling being in a universe billions of years old and still unfolding. The universe isn't just something we study. It's the stage every human story has ever played out on.

So, pause here for a moment. Right at the edge of wonder.

Look up.

And then—let's begin.

CHAPTER ONE: Naming the Vastness

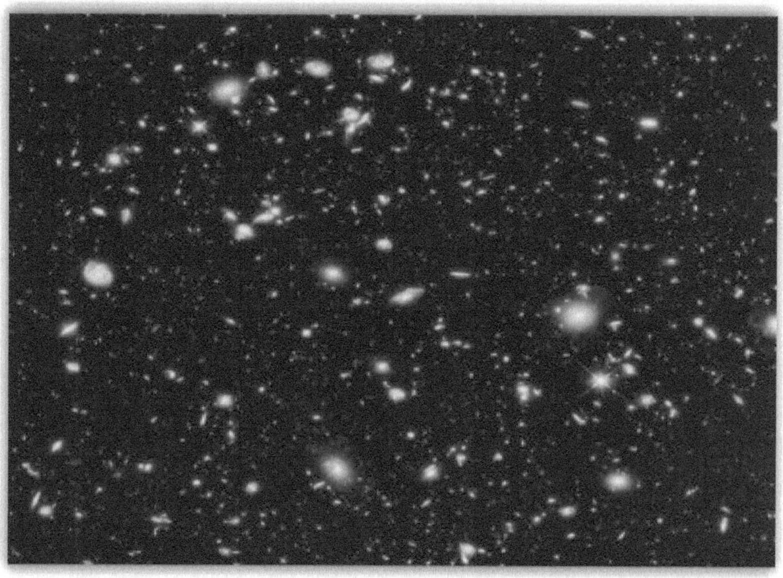

The first thing we do, almost without thinking, when something feels too big, is try to give it a name.

5A kid looks up at the night sky and asks what all that darkness and light actually *is*. Someone a long time ago started drawing invisible lines between stars and telling stories about them. A modern scientist peers through layers of glass and machinery and writes down careful definitions. Different eras, same impulse. Naming is how we get our hands around things that would otherwise slip right through us.

We usually toss around the words *universe* and *cosmos* like they're interchangeable. And, honestly, in everyday conversation they mostly are. But if you slow down and listen to where they came from, they carry different moods with them. *Universe* comes from the idea of everything turned into one thing. All of it, bundled together. *Cosmos*, on the other hand, comes from an older sense of order—of harmony. It hints that what exists isn't just enormous, but arranged somehow. Not chaos shouting into the void, but something with structure. With rhythm.

So, when we use the word *cosmos* here, we're talking about reality as both vast *and* intelligible. Huge, yes but not meaningless.

And that intelligibility is strange, if you really think about it. There's no obvious reason the universe should make sense to us at all. It could have been incomprehensible from top to bottom. And yet, when we look closely enough, patterns show up. The same physical laws seem to work here and billions of light-years away. The same atoms that make up your body light up distant stars. Equations scratched out by human hands describe galaxies that existed long before humans ever did.

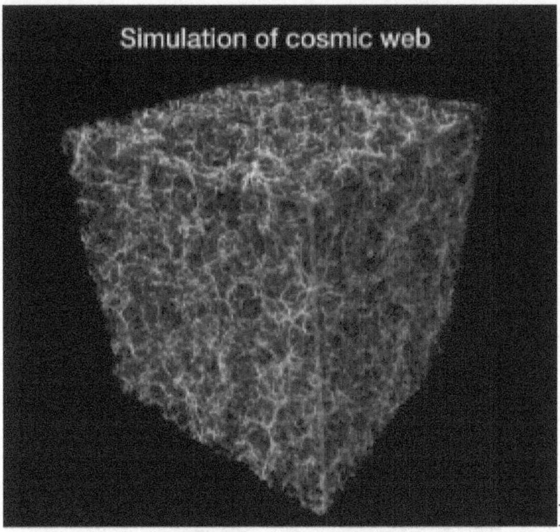

It's easy to miss how odd that is, because we live inside it. Familiar things stop feeling miraculous. But there's nothing inevitable about a universe that can be understood. Ours didn't have to be this way. And yet, it keeps opening itself up to questions. It invites us to look.

There *is* a limit, though—one we should be honest about early on. Everything we know directly about the universe comes from signals that have had time to reach us. Light, radiation, faint whispers traveling across space. That creates a boundary called the *observable universe*. Beyond a certain distance, space has expanded so much that light simply hasn't arrived yet. Those regions might exist. Probably do. But for now, they're beyond our reach.

That distinction matters. When we talk about "the universe," we're often really talking about this visible bubble we're inside. A small window into something potentially far larger. That doesn't make our knowledge useless—but it does keep it humble. Our maps of reality are powerful, but

they are not complete.

Over time, our picture of the cosmos has changed again and again. Early humans put Earth at the center, not because they were arrogant, but because it matched what life felt like. The ground doesn't move. The sky does. It made sense—until it didn't. Slowly, cracks appeared. Earth became just another planet. The Sun became an ordinary star. The Milky Way became one galaxy among hundreds of billions.

Each change was unsettling. Each one pushed us further from the center of the picture. And yet, none of them erased our importance. They just forced us to rethink what importance actually means.

We aren't central in location—but we might be rare in ability. The universe doesn't revolve around us, but it has produced beings capable of understanding parts of it. Meaning doesn't require being in the middle. A single thought can contain a galaxy without sitting at its center.

So, asking what the cosmos it isn't only a question about space, matter, or distance. It's also a question about where we fit into all of this. Are we accidental witnesses in an indifferent expanse, or participants in something still unfolding? Is the universe just a container, or is it a process—one that includes us?

These questions don't weaken science. They exist *because* of it.

As we move on, the ideas will get sharper. Distances will stretch past intuition. Time will start behaving in ways that feel wrong at first. But before all of that, this chapter asks you to hold onto something simple. The cosmos isn't just "out there." It's the deepest background of your own existence.

Every breath you take is a small moment in a story billions of years long. Every thought you have is matter briefly becoming aware of itself. Naming the vastness doesn't shrink it. It reminds us that we already belong to it.

And that realization—that quiet, unsettling, beautiful realization—is where the journey actually begins.

CHAPTER TWO: When Absence Refuses to Stay Empty

Try this for a moment.

Really try it.

Turn everything off.

No light. No sound. No matter. No space to put anything in. No time for anything to happen. Sit with that idea and don't cheat—don't sneak darkness back in, don't imagine a black room or an empty sky.

It falls apart almost immediately.

The mind reaches for darkness, but darkness is already *something*. It has a texture. A shape. A presence. Even our best attempts at imagining nothing quietly turn into *something else*.

Nothingness is surprisingly hard to hold.

For most of human history, "nothing" seemed like the simplest concept imaginable. It was just the absence of things. A blank backdrop. A kind of cosmic silence. Philosophers argued about whether it could even exist, while theologians wondered whether only a divine act could pull something out of it. But the question stayed abstract—interesting, yes, but unreachable. There was no way to test it.

Then physics showed up and ruined the simplicity.

Modern science revealed that what we casually call empty space isn't empty at all. At the smallest scales, fields ripple even when nothing appears to be happening. Energy flickers. Particles pop into existence and vanish again, not because anything pushed them into being, but because the rules of reality allow it. The vacuum isn't quiet. It hums. Absence behaves like it has nerves.

This wasn't poetic interpretation. It was forced on us.

Experiments kept turning up effects that shouldn't exist if "nothing" were truly nothing. Forces acted where there should have been emptiness. Energy appeared without a clear source. The math only worked if space itself had structure. The universe refused to give us a clean, blank slate.

So, the question didn't disappear. It just shifted.

If empty space is already something, then what are we really talking about when we say *nothing*? Is there a deeper layer where even these fields vanish? Or is absolute nothingness an idea the universe simply doesn't allow? Maybe the better question isn't why something exists instead of nothing—but why we ever assumed nothing was an option to begin with.

This is where science slows down.

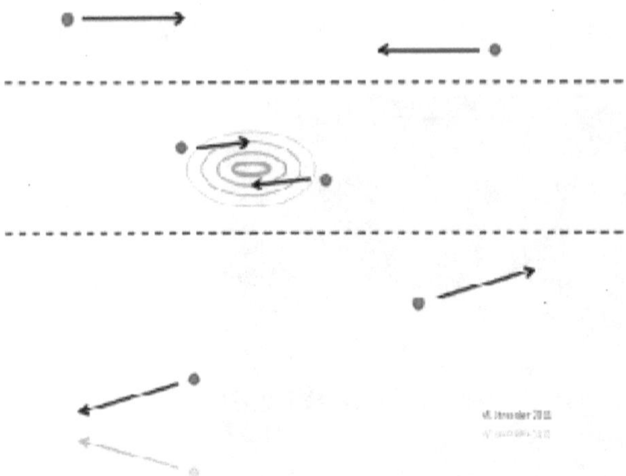

Physics can tell us *how* particles emerge from quantum fields. It can predict probabilities with absurd precision. But it can't yet explain why there is a lawful structure at all—why there are rules that permit fluctuations in the first place. The equations describe behavior, not origin. They tell us what happens, not why reality exists rather than not.

And quietly, philosophy slips back into the room.

Some thinkers suggest existence is necessary—that reality, in some form, cannot fail to exist. Others argue that the laws themselves may be unavoidable, leaving no room for absolute nothingness. Still others imagine our universe as one outcome among many, emerging naturally from a deeper framework we don't yet see.

None of these ideas are settled. And that's the point.

What matters is what they expose about our instincts. Humans tend to treat nothingness as the default and existence as the strange exception. But the universe may be hinting at the opposite. Existence might be the baseline. Order might be more natural than we think. What we call "nothing" could just be a word stretched past its limits.

There's a quieter implication here too—one that hits closer to home.

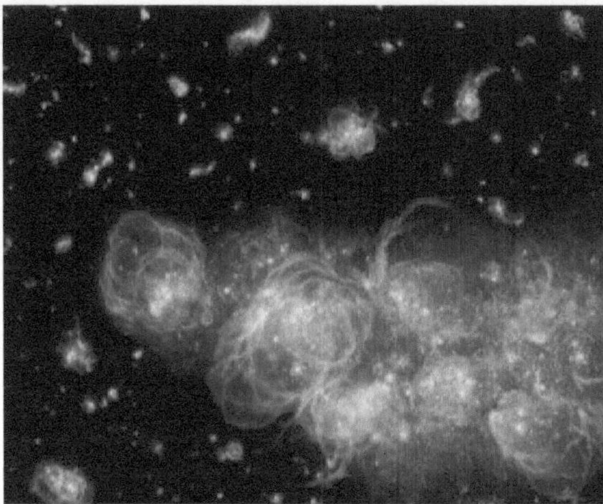

If reality itself resists emptiness, then maybe meaning does as well. We're uneasy with gaps. With silence. With absence. We imagine voids as dangerous or final. But the universe seems structured so that even its emptiest places seethe with possibility.

That's not a scientific claim. It's a reflection—one prompted by what science has uncovered.

The question of nothingness is still open. It may always be. But that's not a failure. It's a reminder that the deepest mysteries aren't always hiding in distant galaxies. Sometimes they're buried inside the most ordinary words we think we understand.

Nothing, it turns out, isn't simple at all.

And if absence itself can't stay empty, then existence may be far more deeply rooted than we ever suspected.

CHAPTER THREE: When Time Learned to Flow

Time feels obvious—right up until you stare at it too hard.

It wakes us up in the morning. It slides shadows along the wall. It quietly carries us from memory to anticipation without asking permission. We chop it into hours and days and years, stick numbers on it, and move on. Most of the time, we never stop to ask what time actually *is*. We just assume it stretches endlessly backward, like a road that's always been there.

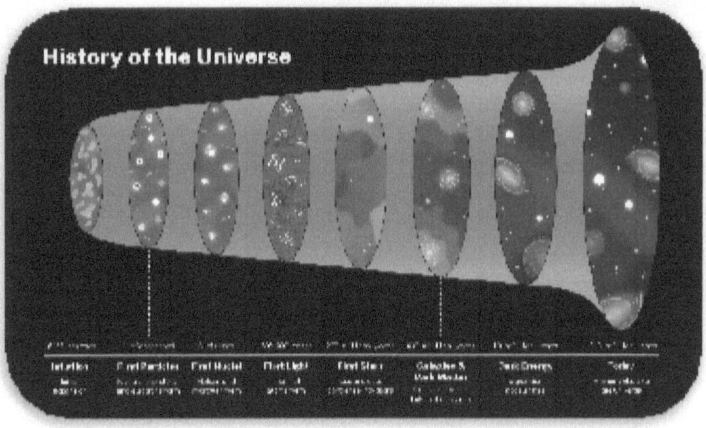

For a long while, no one seriously questioned that assumption. The universe existed. Time existed. Neither seemed to need a beginning.

Then the evidence started to lean in and whisper otherwise.

In the early twentieth century, astronomers noticed something deeply unsettling. Distant galaxies weren't staying put. Their light was stretched toward the red end of the spectrum, as if something were pulling them away from us. And the farther away a galaxy was, the faster it seemed to be receding. This wasn't motion through space the way a car moves down a road. It was space itself expanding.

Once you accept that, a question becomes unavoidable.

If space is expanding now..., what was it like before?

Run the cosmic film backward and the picture changes fast. Galaxies drift closer together. Temperatures climb. Matter compresses. Push far enough, and everything we know gets packed into a state so hot and dense that our current laws of physics begin to fray—and then simply stop working. This limit is what we call the Big Bang, though the name does more harm than good. It wasn't an explosion *in* space. It was the rapid expansion *of* space, happening everywhere at once.

And time, as far as we can tell, is tangled up in that moment.

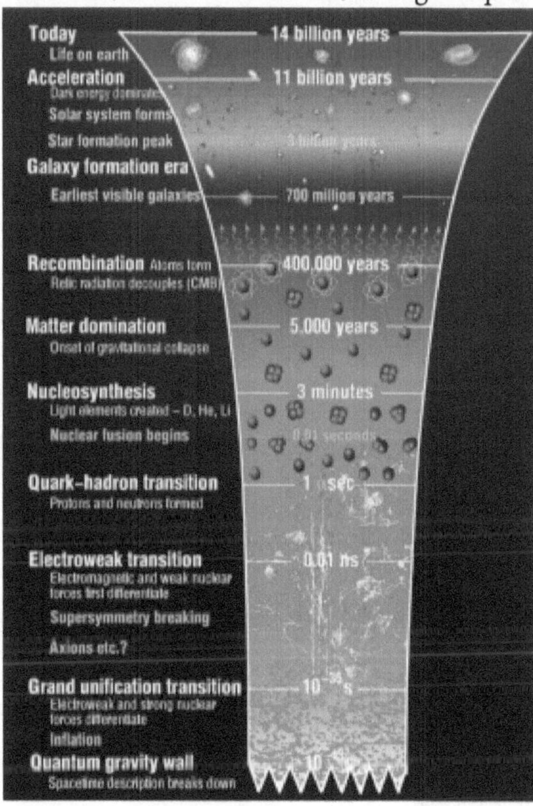

As we approach the earliest instants of the universe, time starts behaving strangely. Within a tiny fraction of a second after the beginning—an interval known as the Planck epoch—our equations lose their grip. Gravity and quantum mechanics refuse to cooperate. Cause and effect blur together. Asking what came before this point may be like asking what lies north of the North Pole. The question assumes a direction that no longer exists.

This isn't metaphor for effect. It's a conclusion forced on us by math and observation.

Understandably, many people find this deeply uncomfortable. We're used to thinking of time as a container—something events happen *inside*. The Big Bang suggests something far more radical. Time itself may be part of the universe, emerging alongside space and matter rather than existing beforehand. There may not have been a "before" at all, at least not in any way that makes sense to us.

Still, cosmology doesn't stop just because things get strange.

Scientists have proposed models that try to peer beyond this boundary. Some suggest a brief period of rapid inflation—an extraordinary expansion driven by unknown physics that shaped the universe we see today. Others explore bouncing universes, where a previous contraction gave rise to our expansion. Still others imagine eternal frameworks in which universes form again and again.

These ideas are ambitious. Mathematical. Incomplete. Some of them will almost certainly turn out to be wrong. But they all come from the same stubborn impulse—the refusal to stop asking just because the answers are hard.

There's a subtle shift that happens when you accept that time may have a beginning. Existence becomes finite in one direction, but not fragile. The universe didn't pop out of nothing in a casual sense. It emerged from a boundary we don't yet understand. That distinction matters. It replaces tidy stories with honest uncertainty.

It also changes how we see the present.

If time had a beginning, then every moment since has been part of a single, continuous unfolding. The stars forming tonight, the thoughts forming in your mind, the ancient light landing on your eyes from distant galaxies—all of it traces back to the same origin. The universe isn't a static stage. It's a story in motion, and time is the medium that carries it forward.

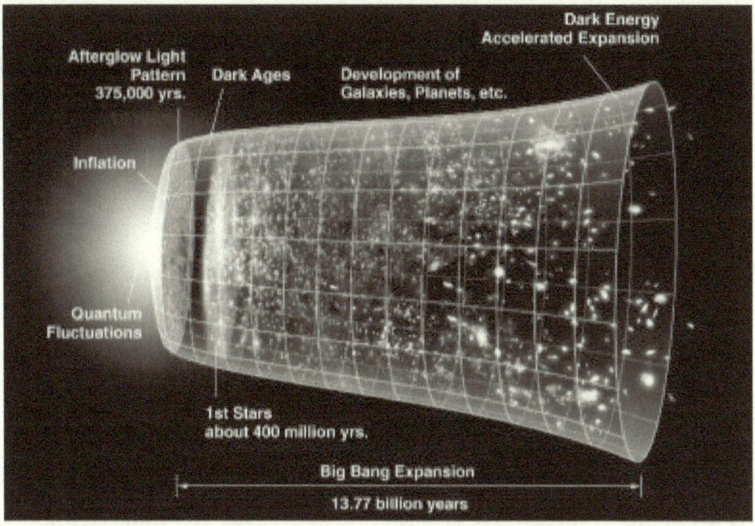

We often imagine eternity as endless duration. Cosmology hints at something subtler. Eternity may not mean infinite time, but whatever depth of reality lies beneath time itself. Science can't define that depth. It can only show us where its tools go quiet.

And that quiet matters.

At the beginning of time, our equations fall silent. Our intuitions fail. Language strains. But this isn't defeat. It's a reminder that reality is larger than our current understanding. The edge of time isn't a wall—it's a horizon.

And horizons are meant to be approached.

The universe has been expanding and cooling for billions of years, carrying time along with it. In the next chapter, we'll follow that motion outward and ask what it means for a universe that refuses to stand still.

CHAPTER FOUR: A Universe That Refuses to Hold Still

The universe isn't a finished thing.

It's not a completed structure sitting there, polished and permanent. It's something that's still happening. Still moving. Still becoming.

For most of human history, though, it didn't look that way. The cosmos seemed grand but fixed, like a vast architectural ceiling stretched over the world. The stars held their positions. The heavens felt eternal. Change belonged to Earth—weather, seasons, birth and decay—not to the sky itself. That picture made sense. Human lives are short, and the universe moves slowly. Only observations made patiently, across generations, revealed that the apparent stillness was misleading.

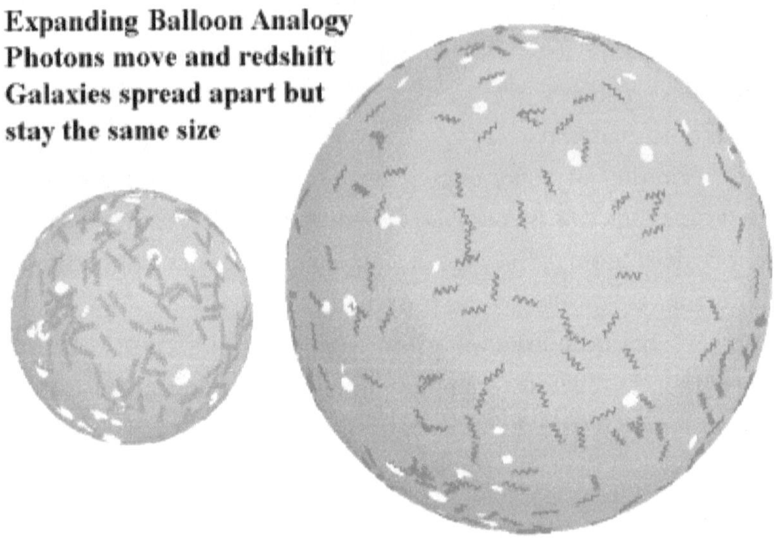

When astronomers finally began measuring the light from distant galaxies with enough precision, something unsettling appeared. The light arriving from faraway systems was stretched, pulled toward redder wavelengths. This wasn't random. It wasn't confined to one direction or one corner of the sky. It showed up everywhere, as if the entire universe were participating in a single motion.

There was no way around it.

Space itself is expanding.

This expansion doesn't behave the way explosions do in everyday life. There's no center point where everything started flying outward, and no edge where the universe ends. Every region of space moves away from every other region—not because galaxies are racing through space, but because the space between them is stretching. A common image compares this to dots drawn on the surface of a balloon. As the balloon inflates, each dot moves away from the others, yet none of them sits at the center. The motion belongs to the surface itself.

Once that idea sinks in, it quietly rearranges everything.

If space is expanding now, then it must have been smaller before. Distances that span millions of light-years today were once compressed into far tighter volumes. The universe was hotter. Denser. Stranger. The early cosmos wasn't just a scaled-down version of what we see now—it was a fundamentally different environment, one where processes occurred that can no longer happen naturally.

Expansion also changes how we see.

Because light travels at a finite speed, looking deep into space means looking back in time. When we observe a galaxy billions of light-years away, we're not seeing it as it is *now*. We're seeing it as it was billions of years ago. The universe becomes a kind of layered archive, with distance doubling as history. The farther we look, the earlier the chapter.

But even that record has limits.

Beyond a certain distance, space has expanded so quickly that light from those regions will never reach us. This boundary is called the cosmic horizon. It isn't a wall you could bump into. It's a limit created by expansion itself. Events beyond it are real, but forever inaccessible to direct observation.

That realization carries a quiet weight.

The universe may be far larger than what we can ever see—possibly infinite. Our observable region could be just a small patch in something vastly bigger. Science doesn't accept this idea because it's comforting. It accepts it because the math leaves little choice. Once again, knowledge grows, and certainty thins.

Expansion also points forward, not just backward.

If space continues to stretch, galaxies will drift farther apart. Over unimaginable spans of time, distant systems will fade from view, their light stretched until it disappears entirely. The night sky of the far future may look emptier than ours—not because stars are gone, but because the universe has carried them out of reach.

Still, expansion isn't only a story of loss.

It's also the reason anything exists at all. Without expansion, matter would have collapsed too quickly. Without gravity, it would have spread too thin. The universe lives in a narrow balance between these tendencies, one that allows structure to form—galaxies, stars, planets, and eventually life.

What's most surprising is that this expansion isn't slowing down.

Observations suggest it's accelerating; driven by something we call dark energy—a name that admits how little we truly understand it. Space isn't just stretching. It's doing so faster and faster, as if growth itself is written into the universe's deepest rules.

So, this is the picture we're left with.

A universe that is restless at every scale. Dynamic. Unfinished. Nothing about it suggests permanence or finality. Even the largest structures we know are temporary arrangements in a much longer unfolding.

To live in such a universe is to live inside a process, not a monument.

Every moment exists because space continues to expand, time continues to flow, and the laws of physics continue to hold—quietly, reliably. The universe doesn't pause while we try to understand it. It moves on, carrying our questions with it.

And now, to understand where those questions truly begin, we have to look back to an earlier era—-after expansion had started, but before stars had ignited. A time when the universe was dark, simple, and filled with the raw ingredients of everything that would one day emerge.

CHAPTER FIVE: From Fire to Form

Long before the first star ever flickered to life, the universe existed in a state that was almost brutally simple.

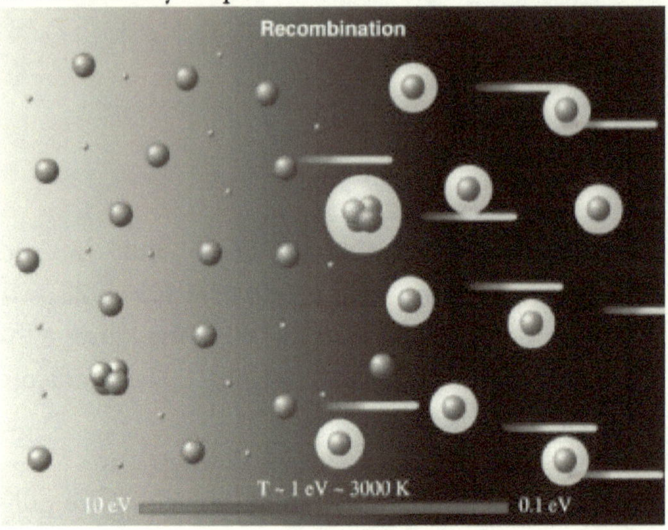

Right after expansion began, everything was packed together—energy, particles, light—all locked in a constant, chaotic exchange. Photons couldn't travel anywhere without crashing into charged matter. Light existed, sure, but it couldn't *go* anywhere. No images. No distance. No structure. Just a hot, blinding fog where nothing could stand apart from anything else.

And then, slowly—almost anticlimactically—things began to cool.

As space stretched, temperatures dropped. Energy thinned out. Eventually, the universe crossed a quiet threshold. Protons finally held onto electrons. Neutral atoms formed. And with that one subtle change, light slipped free.

That light never vanished.

It's still here, washing over the universe from every direction as a faint, microwave afterglow—the cosmic microwave background. It's the oldest light we can see, a snapshot from when the universe was only a few hundred thousand years old. When scientists study it, they aren't seeing stars or galaxies. They're seeing the universe itself at the exact moment it became transparent.

What's striking isn't just how old this light is—it's how imperfect it is.

At first glance, it's almost uniform. But "almost" matters. Tiny temperature differences ripple across it, variations so small they're measured in fractions of a degree. They seem trivial. They aren't. Without them, the universe would be empty and boring forever.

Perfect smoothness produces nothing.

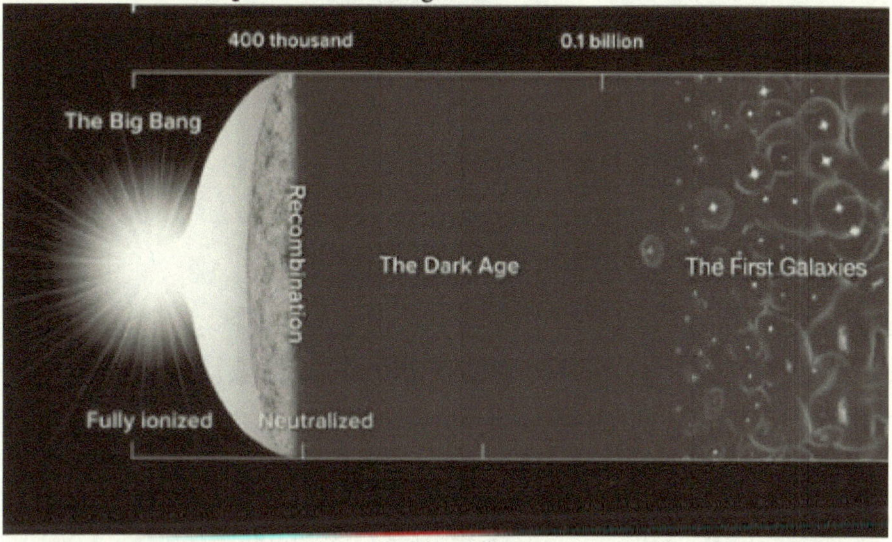

Those faint irregularities were seeds. Slightly denser regions tugged just a bit harder on their surroundings. Gravity, patient and unhurried, amplified those differences over millions of years. Matter drifted inward. Structures began to take shape—not suddenly, not dramatically, but inevitably.

Between the release of that ancient light and the birth of the first stars lies a stretch of time we call the cosmic dark ages. The name sounds empty, but the era wasn't. The universe was filled with neutral hydrogen gas, spread thinly through expanding space. No stars burned. No galaxies glowed. Darkness ruled—but motion never stopped.

Gravity doesn't need light.

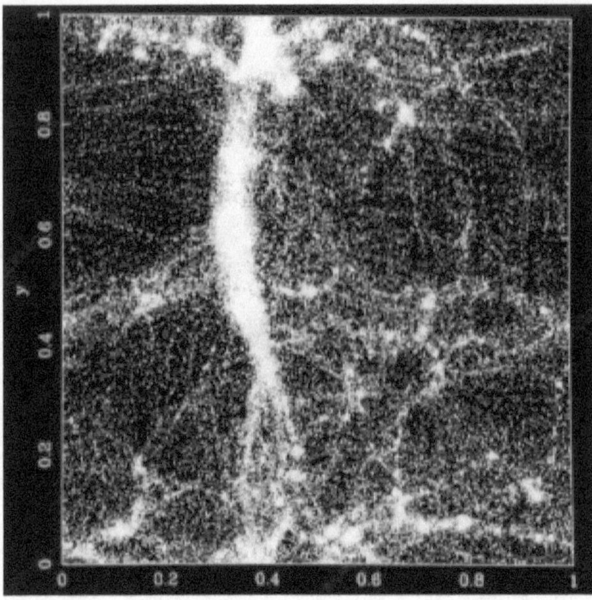

Gas slowly slid toward denser regions. Clouds thickened. Their centers warmed. Pressure crept upward. Nothing rushed, yet everything moved toward a tipping point—a moment when collapse would ignite something entirely new.

This was the universe's quiet transition from simplicity to structure.

No guiding hand rearranged matter step by step. The same laws applied everywhere, all the time. Gravity pulled. Expansion pushed back. From that tension, form emerged naturally. The universe didn't need a plan. It only needed consistency and time.

That idea is unsettling—and profound.

Complexity didn't appear *despite* simplicity. It emerged *because* of it. With only a few ingredients and a handful of rules, the universe produced galaxies, chemistry, and eventually life itself. Not instantly. Not deliberately. But reliably.

Creation, in this sense, was never a single moment. It was—and still is—a process. The early universe didn't deliver finished structures. It laid foundations. It made possibilities inevitable.

So, when the first stars finally ignited, they weren't isolated miracles. They formed within a cosmos already shaped by invisible patterns, molded by forces acting since the first light broke free.

Darkness ended—but mystery didn't.

The universe had taken its first real steps from fire toward form.

In the next chapter, those early structures will grow vast and interconnected, as gravity builds immense cities of stars linked across cosmic distances—revealing that even on the largest scales, the universe carries a hidden architecture.

CHAPTER SIX: The Invisible Skeleton of the Cosmos

If you could step far enough back from the universe, far beyond the scale of individual galaxies, what you would see would not be random scatter or uniform haze, but something unexpectedly familiar. The cosmos, viewed on its largest scales, resembles a web. Long filaments stretch across space, intersecting at dense nodes, enclosing vast empty regions like bubbles frozen in mid expansion. It looks less like chaos and more like architecture.

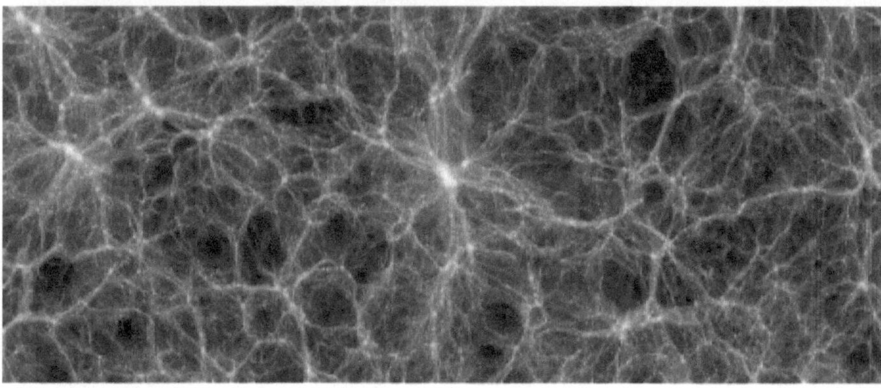

This structure was not designed in advance. It emerged.

As gravity continued its slow work after the cosmic dark ages, matter did not collapse evenly. Regions seeded by early density variations drew in surrounding material, growing heavier and more influential over time. Matter streamed inward along preferred paths, guided by gravity and resisted by expansion. Over billions of years, this motion carved out the large-scale structure of the universe, a sprawling network known today as the cosmic web.

Galaxies are not scattered randomly through space. They live in neighborhoods.

Along the filaments of this web, gas and dark matter flow together, feeding the growth of galaxies and galaxy clusters. At the intersections, matter piles up into enormous concentrations, forming clusters that contain hundreds or even thousands of galaxies bound together by gravity. Between these structures lie cosmic voids, immense regions with very little matter at all, places where gravity lost the contest early and emptiness prevailed.

These voids are not failures of creation. They are part of the pattern.

What is striking is how early this structure began to form. The blueprint was present in the tiny fluctuations seen in the cosmic microwave background. Gravity simply amplified what was already there. The universe did not need to invent complexity later. It carried the seeds of structure from its earliest moments, waiting patiently for time to do the rest.

Within this grand framework, individual galaxies began to take shape.

As gas fell into dark matter halos, it spun, cooled, and flattened into rotating disks or gathered into glowing ellipsoids. Stars ignited within these systems, lighting up the cosmic web from the inside. Spiral galaxies formed graceful arms rich with gas and ongoing star formation. Elliptical galaxies emerged more quietly, dominated by older stars and shaped by ancient mergers. Each galaxy became a long-lived island of light embedded in a much larger gravitational environment.

Our own home exists within this architecture.

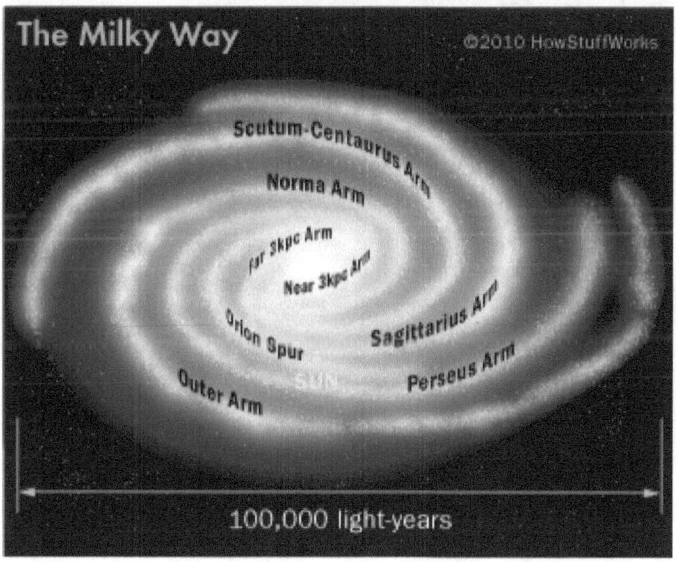

The Milky Way is part of a small collection known as the local group, a modest gathering of galaxies bound together by mutual gravity. This group sits along a filament that connects to larger clusters, which themselves trace the contours of the cosmic web. Even our seemingly isolated night sky is part of a much greater arrangement, extending far beyond what the eye can see.

There is a subtle lesson here about perspective.

From Earth, the universe appears crowded with stars. From the scale of galaxies, space seems vast and empty. From the scale of the cosmic web, structure reemerges. Meaning depends on distance. What appears insignificant at one level becomes essential at another. The universe teaches this lesson repeatedly, as if insisting that no single viewpoint is sufficient.

This architecture also tells us something about time.

Galaxies are not permanent fixtures. They collide, merge, grow, and sometimes fade. Clusters evolve. Filaments shift. The cosmic web itself is dynamic, slowly stretching as the universe expands. Even the largest structures known are temporary arrangements in a much longer story. Stability exists, but only on human timescales.

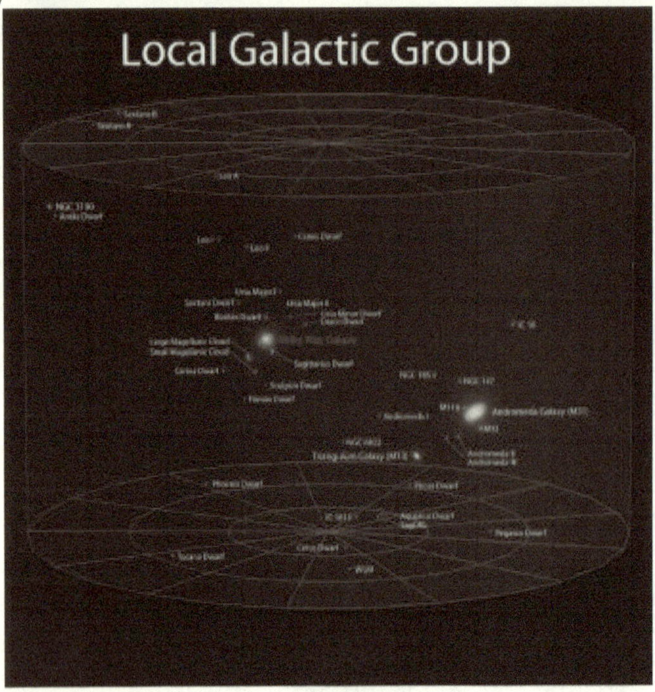

And yet, within this impermanence, something remarkable occurs. Galaxies create environments where complexity can flourish. Stars form. Elements accumulate. Planets assemble. Life becomes possible. The vast architecture of the universe provides the conditions for intimate stories to unfold.

The universe builds cities of stars not for display, but as a natural outcome of its laws.

Standing beneath the night sky, it is easy to feel detached from these immense structures. But every photon that reaches your eyes has traveled through this cosmic architecture to find you. Every atom in your body has passed through multiple generations of stars formed within it. You are not merely observing the cosmic web. You are one of its consequences.

The universe is large beyond comprehension, yet it is not hollow. It is shaped, connected, and quietly ordered. On its grandest scales, it reveals that complexity is not an exception, but an expectation.

And within those vast cities of stars, a new kind of creation begins, as individual suns ignite and take on lives of their own, becoming the furnaces where matter itself is transformed.

CHAPTER SEVEN: The Long Lives of Suns

Stars don't start out as neat pinpricks of light. That part comes much later.

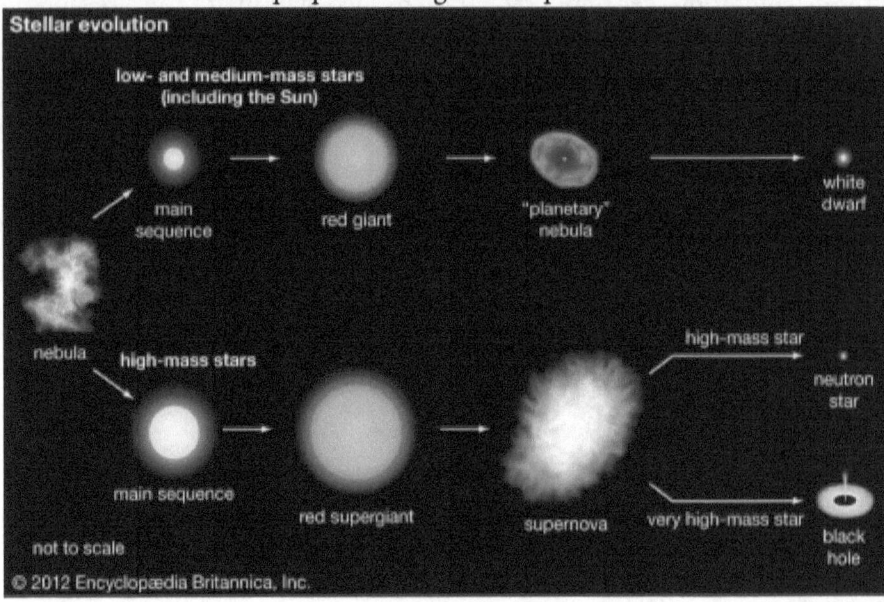

At first, they're almost nothing—thin, drifting clouds of gas and dust stretched across absurd distances, hanging in space with no urgency at all. For millions of years, nothing seems to happen. Then gravity nudges things just enough. The cloud buckles, fragments, and slowly pulls itself inward. It's quiet work. Patient work.

As a fragment collapses, it warms. Pressure builds. Somewhere deep inside, the conditions cross a line that physics has been waiting for. Hydrogen nuclei begin to fuse. Not dramatically—at first—but decisively. Energy pushes outward. Gravity pulls inward. The two forces settle into an uneasy truce.

A star exists.

Once this ignition happens, the star enters the longest chapter of its life. Most stars spend the overwhelming majority of their time here, calmly converting hydrogen into helium, shining steadily into the dark. This balance isn't fixed or guaranteed. It's constantly maintained, moment by moment, like standing still while two invisible hands push from opposite sides.

Mass decides almost everything.

Small stars sip their fuel slowly and can endure for trillions of years, longer than the universe has existed so far. Massive stars burn like they're in a hurry, blazing brilliantly and dying young after only a few million years. Brighter doesn't mean longer-lived. If anything, it means the opposite.

But light isn't the most important thing stars do.

Inside their cores, stars are factories. Through nuclear fusion, simple elements are reshaped into more complex ones—carbon, oxygen, nitrogen, and others that chemistry and biology depend on. The universe begins with little more than hydrogen and helium. Stars do the hard work of making everything else possible.

Eventually, though, the balance breaks.

When a star runs out of usable fuel in its core, gravity stops negotiating. What happens next depends—again—on mass. Smaller stars release their outer layers gently, leaving behind white dwarfs: dense, hot embers slowly fading over unimaginable time. Larger stars end far less politely. Their cores collapse in an instant, rebounding in supernova explosions that can briefly outshine entire galaxies.

These explosions are not just endings. They are distributions.

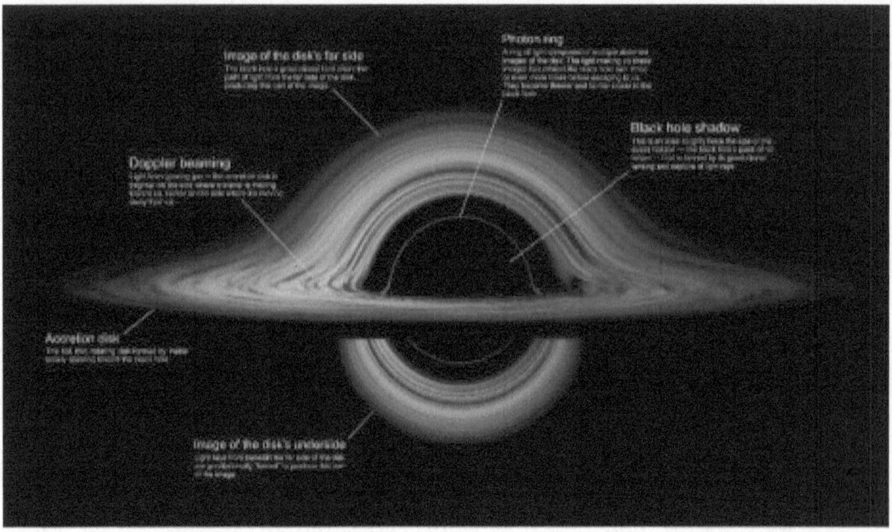

Supernovae scatter newly forged elements across space, seeding surrounding regions with the raw material for future stars, planets, and chemistry. They also create the heaviest elements—gold, uranium—under conditions so extreme they occur almost nowhere else. Destruction and creation happen in the same breath.

Sometimes, something even stranger remains.

A collapsed core may become a neutron star, so dense that matter itself is pushed to its limits—a teaspoon would outweigh a mountain. In rarer cases, gravity wins completely, forming a black hole were space and time curve inward beyond ordinary description. These objects aren't mistakes or cosmic glitches. They are lawful outcomes, just as natural as ignition.

One famous reminder of this process still lingers in our sky: the **Crab Nebula**, the expanding debris of a star whose death was recorded by human observers nearly a thousand years ago.

This cycle repeats everywhere.

Stars form, endure, and disappear across the universe in rhythms shaped by physics and time. Their lives dwarf human history, yet barely register against cosmic scales. They are dependable and fragile at once. Without them, the universe would remain dark, simple, and sterile.

There's something quietly personal in all of this.

Every atom in your body heavier than hydrogen was forged inside a star that lived and died before Earth ever existed. The calcium in your bones, the iron in your blood, the oxygen you breathe—all of it passed through stellar interiors long ago. This isn't poetry. It's chemistry.

Stars aren't decorations.

They are ancestors.

And as the story continues, their remnants will push physics to its breaking point, forcing us to question what space, time, and even reality itself truly is.

CHAPTER EIGHT: Where Gravity Wins

Not all endings are quiet, and not all remnants are gentle.

When the most massive stars reach the limits of what nuclear fusion can sustain, their collapse does not slow or settle. Gravity tightens its grip without compromise, crushing the star's core beyond any familiar state of matter. What forms is not simply an object, but a region where space and time themselves are bent so severely that ordinary descriptions begin to unravel.

This is what we call a black hole.

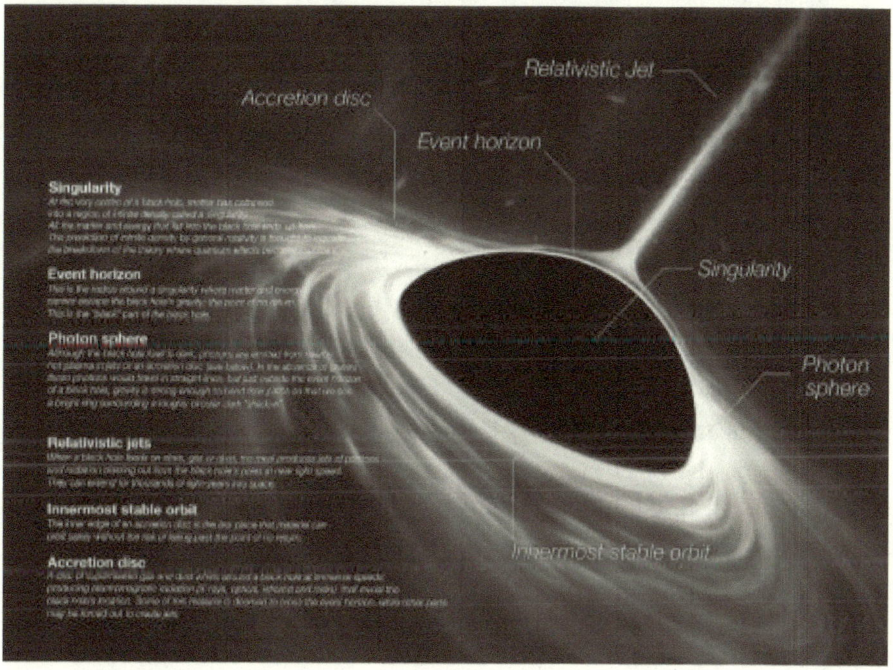

At its boundary lies the event horizon—a one-way surface in spacetime. Beyond it, nothing that happens can influence the outside universe. Light emitted inside cannot escape. Signals fall inward and never return. To distant observers, the black hole appears dark, not because it is empty, but because gravity has become absolute.

Yet black holes are often misunderstood.

They are not cosmic vacuum cleaners tearing through space. From a distance, a black hole behaves just like any other object of equal mass. Stars can orbit it calmly. Gas clouds can drift past untouched. Only when matter strays too close does' gravity seizes it irreversibly. What makes black holes extraordinary is not aggression, but finality.

When matter does fall inward, it rarely does so quietly.

Gas spiraling toward a black hole forms an accretion disk, flattening and heating as friction and gravity accelerate particles to extreme speeds. The disk glows intensely, releasing enormous amounts of energy. In some systems, black holes also launch narrow jets of matter and radiation that shoot outward at nearly the speed of light, extending thousands of light-years and reshaping entire regions of space. Far from being inert voids, black holes can dominate their environments.

At the heart of almost every large galaxy lies one.

Supermassive black holes contain the mass of millions or even billions of suns compressed into volumes smaller than our solar system. Their origins are still debated, but their influence is undeniable. They regulate star formation, sculpt galactic structure, and anchor the motion of stars on vast scales. At the center of our own galaxy resides **Sagittarius A***, an invisible presence revealed only through the dance of stars orbiting an unseen mass.

Black holes force physics into unfamiliar territory.

According to general relativity, matter falling beyond the event horizon continues inward toward a singularity—a point where density becomes infinite and the equations describing spacetime collapse under their own assumptions. Space and time lose their usual meanings. Prediction gives way to symbols that gesture toward something real, but inaccessible.

This breakdown is not a failure of science. It is a signpost.

Black holes sit at the intersection of gravity and quantum mechanics, exposing the incompatibility of our most successful theories. Efforts to reconcile them have reshaped modern physics, introducing ideas that suggest information itself may be as fundamental as matter or energy.

One of the most unsettling discoveries is that black holes are not entirely black.

Quantum effects near the event horizon allow them to emit faint radiation, slowly losing mass over immense spans of time. This process implies that black holes can eventually evaporate. With that realization comes a deep paradox: if a black hole disappears, what happens to the information about everything that fell into it? Does the universe forget, or is the information preserved in some hidden form?

The question reaches beyond black holes.

Physics has long relied on the assumption that information is never destroyed. If that principle fails here, it fails everywhere. The debate has transformed how scientists think about reality itself, turning black holes into laboratories for ideas about time, causality, and the structure of the universe.

There is a strange symmetry at work.

Stars collapse to form black holes, and black holes influence the birth and evolution of stars. Creation and destruction are not opposites, but phases of a single cycle. The universe tests its own laws through these extremes, pushing structure to its limits and watching what survives.

To contemplate a black hole is to stand at the edge of understanding.

We can predict their motions, map their surroundings, and measure their effects with extraordinary precision. And yet, at their core lies a mystery that resists every attempt at simplification. Black holes remind us that the universe is not obligated to be fully understood—even as it invites us to keep trying.

CHAPTER NINE: The Universe We Cannot See

By the time astronomers learned how galaxies truly move, it became obvious that something was missing.

Stars in spiral galaxies orbit their centers in ways that simply should not work. According to everything we understand about gravity and visible matter, stars farther from the galactic core ought to move more slowly, just as distant planets drift more gently around the Sun. But that isn't what happens. The outer stars race along just as fast as those near the center, as if held in place by an unseen mass.

At first, the problem felt technical.

Maybe distances were misjudged. Maybe telescopes were lying. Maybe gravity itself needed revision. But as observations accumulated—from individual galaxies to massive clusters—the same pattern appeared again and again. There was more mass present than could be accounted for by anything emitting light.

Something invisible was shaping the universe.

This unseen component came to be called dark matter.

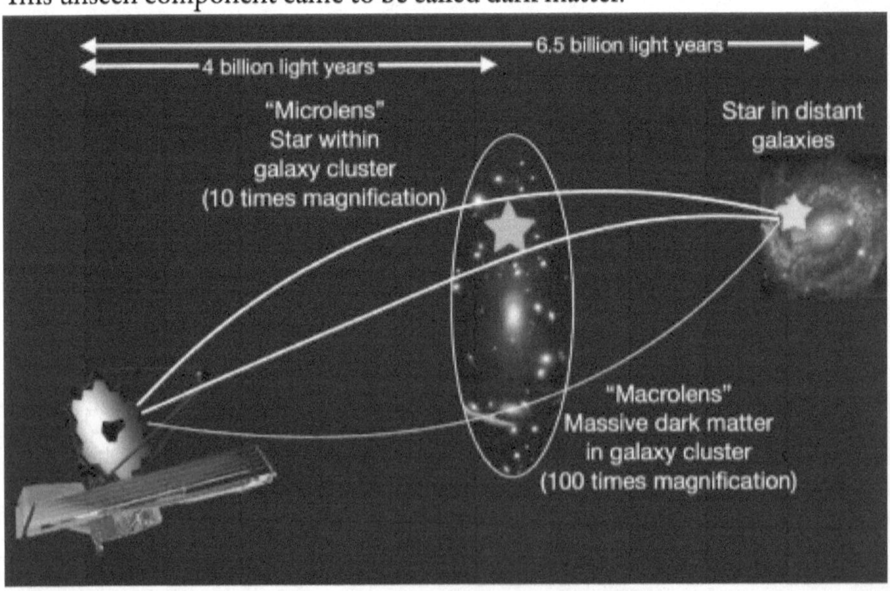

Dark matter does not glow, absorb, or reflect light. It does not form stars, planets, or clouds. It passes through ordinary matter almost without interaction.

And yet its gravitational pull is unmistakable. Without it, galaxies would tear themselves apart. Clusters would never assemble. The large-scale structure of the universe—the vast filaments and voids—would not exist.

Its presence is revealed only indirectly.

When light from distant galaxies passes near massive clusters, gravity bends its path, distorting images into arcs and smears in a phenomenon known as gravitational lensing. By measuring how much the light bends, astronomers can calculate the mass doing the bending. Time and again, the numbers refuse to match what we can see. Most of the mass is invisible.

On cosmic scales, dark matter outweighs ordinary matter by more than five to one.

This realization quietly rewrote our understanding of existence. Everything familiar—stars, planets, people—turns out to be a cosmic afterthought. We are made of a rare kind of matter, glowing briefly within a universe dominated by something we cannot touch or directly observe.

And then came a second shock.

As astronomers refined measurements of the universe's expansion, they expected gravity to be slowing it down. After all, matter pulls inward. Expansion should decelerate. Instead, distant galaxies revealed the opposite. The expansion of the universe is accelerating. Space itself is stretching faster with time, as if driven by a pressure built into reality.

This influence is called dark energy.

Unlike dark matter, which clumps and pulls, dark energy appears smooth and uniform. It does not gather into structures. It fills space evenly, pushing everything apart. Its nature remains unknown. It may be a property of empty space itself. It may point to new physics beyond our theories. Or it may reflect a deep misunderstanding of gravity on the largest scales.

What *is* clear is its dominance.

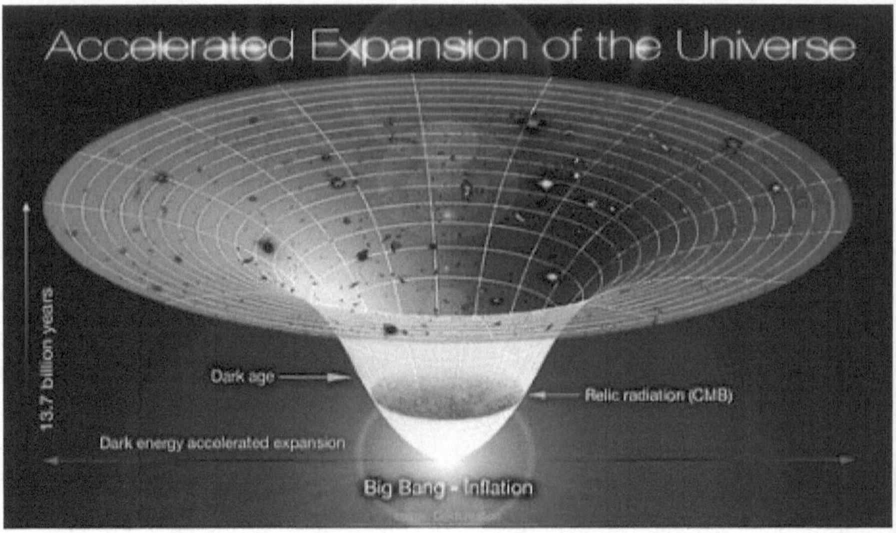

Dark energy makes up most of the universe's total energy content. It governs the long-term fate of cosmic expansion. Over immense spans of time, it will determine which galaxies remain visible, which drift beyond reach, and whether structure itself can endure. The future of the cosmos is controlled by something we do not yet understand.

There is something profoundly humbling in this picture.

Human knowledge has advanced far enough to map galaxies billions of light-years away, yet the forces driving their motion remain largely mysterious. We can describe what dark matter and dark energy *do* with astonishing precision, but not what they *are*. Our equations work, even as our intuition fails.

This is not a weakness of science. It is a sign of its integrity.

Science does not pretend mystery is understanding. Names like "dark matter" and "dark energy" are not explanations—they are markers. They draw a boundary around ignorance and invite further exploration. History suggests such boundaries rarely last forever, though crossing them may take generations.

There is a deeper lesson hidden here.

The universe is not structured around what is most visible. Its foundations are concealed. What we see depends on what we cannot see. Motion is shaped by unseen mass. Expansion is driven by invisible pressure. Reality is layered, with the surface resting on deeper, hidden dynamics.

In that way, the cosmos feels strangely familiar.

Much of what shapes human life—beliefs, memories, histories, emotions—remains unseen, yet profoundly influential. The universe operates by a similar logic. The visible emerges from the invisible, and understanding requires patience, humility, and imagination.

CHAPTER TEN: Rules Written into Reality

Before talking about particles or forces, it's worth stopping to notice something almost invisible because it's so constant: the universe behaves reliably. Things don't just happen at random from one moment to the next. Causes lead to effects. Experiments give the same results tomorrow as they did yesterday. Patterns persist. Without this quiet regularity, science—and ordinary life—would collapse instantly.

What we call the laws of physics are not laws in the human sense.

They aren't commands issued to the universe, nor agreements it somehow signed. They are descriptions. When physicists write equations, they aren't telling nature what to do; they're translating what nature has *always* done into mathematical language. The laws don't prescribe behavior. They reveal it.

What's remarkable is how much these laws accomplish with so little.

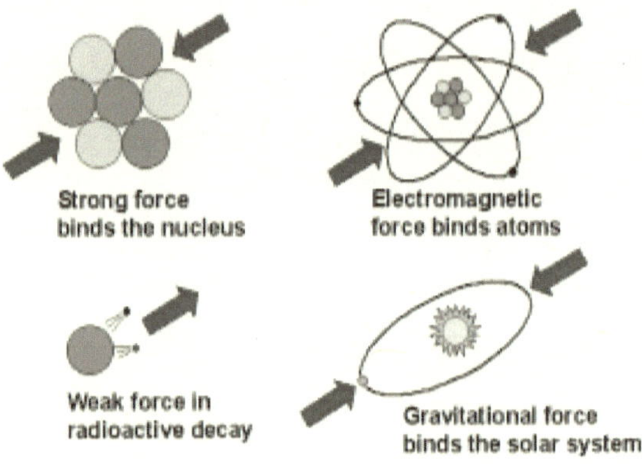

At the deepest level, nearly all physical phenomena arise from a small set of interactions. Gravity governs the motion of planets and galaxies and shapes spacetime itself. Electromagnetism controls light, chemistry, and the structure of atoms. Two nuclear forces operate at subatomic scales, binding matter together and powering stars. From this short list emerges everything we see—from thunderstorms to living cells.

That simplicity is misleading.

Beneath everyday experience lies the quantum world, where intuition breaks down. Particles behave like probabilities instead of objects. They can exist in multiple states at once, influence one another across distance, and appear or vanish according to strict statistical rules. Reality at its foundation is not built from tiny solid pieces, but from abstract relationships governed by precise mathematics.

One idea quietly ties all of this together: symmetry.

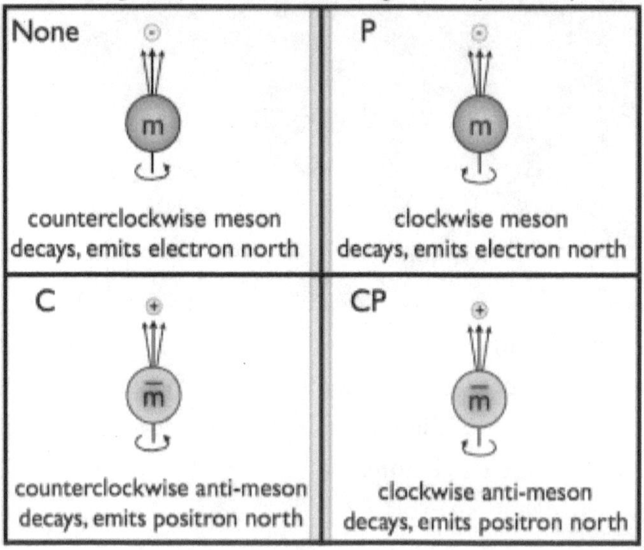

In physics, symmetry doesn't just mean visual balance. It means invariance—the fact that certain properties remain unchanged under transformation. The laws of physics don't depend on where you are, when you run an experiment, or which direction you face. From these symmetries flow conservation laws, like the conservation of energy and momentum. Stability emerges because reality treats equivalent situations the same way.

This is not philosophical decoration. It's exact.

The deeper physicists dig, the more they find that order is not imposed afterward. It is built into the structure of the equations themselves. Many of the most successful theories—such as the **Standard Model** and **General Relativity**—were guided as much by mathematical elegance as by experiment. Beauty turned out to be a reliable clue.

That leads to an unsettling question.

Why does mathematics work at all? Numbers and symbols are inventions of the human mind, yet they describe physical reality with eerie precision. Abstract structures developed without reference to nature end up governing atoms, stars, and spacetime. The universe doesn't just follow rules—it follows rules that can be written down, shared, and understood.

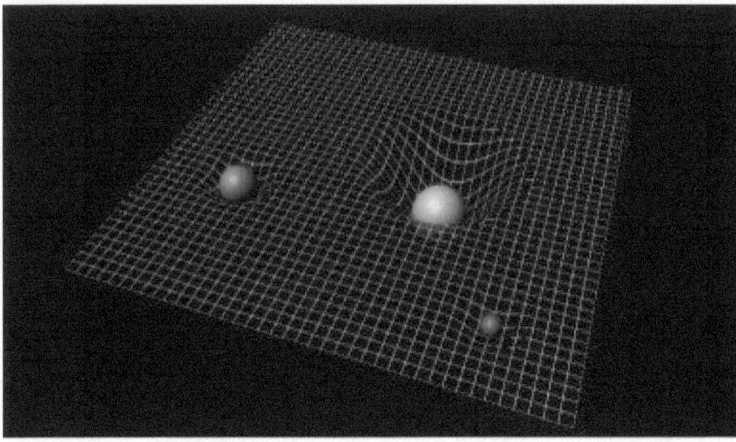

Some argue this is inevitable. We design mathematics to describe patterns, so of course it fits. Others suspect something deeper—that reality itself is mathematical in nature, and human reasoning happens to align with it. Science doesn't resolve this tension, but it makes it impossible to ignore.

There are limits to what laws can explain.

Equations tell us how systems evolve once they exist. They do not explain why these particular laws exist rather than others, or why they allow complexity, stability, and consciousness at all. Physics can trace the universe's behavior with extraordinary accuracy, but the origin of the laws themselves remains unexplained. They are discovered, not derived.

That doesn't weaken them. It sharpens the mystery.

The universe behaves as if it is written in a language older than humanity, yet open to comprehension. The same laws that governed the earliest moments of cosmic history are operating inside your body right now. They have never changed. They do not adapt. They simply *are.*

To live in such a universe is to live inside reliability itself.

The laws of physics do not promise meaning or comfort. They offer consistency. And from that consistency arises everything else—including the possibility of wondering why anything exists at all.

CHAPTER ELEVEN: A Delicate Balance That Did Not Have to Be

As physicists pushed deeper into the structure of reality, an unsettling realization emerged—not from philosophy, but from calculation. The universe appears to rest on a razor's edge. Adjust certain numbers by even the smallest amount, and everything familiar dissolves. Stars never ignite. Chemistry fails. Matter never organizes into anything more complex than a thin, drifting fog.

These numbers are known as fundamental constants.

They are not derived from deeper principles—at least not yet. They are measured. The strength of gravity. The charge of the electron. The masses of particles. The rate at which space expands. Together, they determine how reality unfolds. What startled scientists was not that such constants exist, but how narrowly constrained they appear to be.

If gravity were just a little stronger, stars would burn too fast and collapse violently, leaving no time for stable planets. If it were slightly weaker, matter would never clump at all. If the electromagnetic force were altered even modestly, atoms would fall apart or refuse to form. If the early expansion rate of the universe differed by a tiny fraction, space would either tear itself apart too quickly or collapse back on itself before structure could emerge.

The margin for error is astonishingly small.

This observation has nothing to do with human preference. It follows directly from physics. Complex structures require long-lived stars, stable atoms, and persistent energy gradients. Those conditions exist only within a narrow range of parameters. Outside that range, the universe might still exist—but it would be sterile, simple, and silent.

This leads to a question that refuses to go away.

Why do the constants fall within this narrow, life-permitting window at all? Why does the universe possess precisely the structure that allows complexity to arise? Is this an extraordinary coincidence, a necessity we do not yet understand, or something else entirely?

One response is often summarized by the anthropic principle.

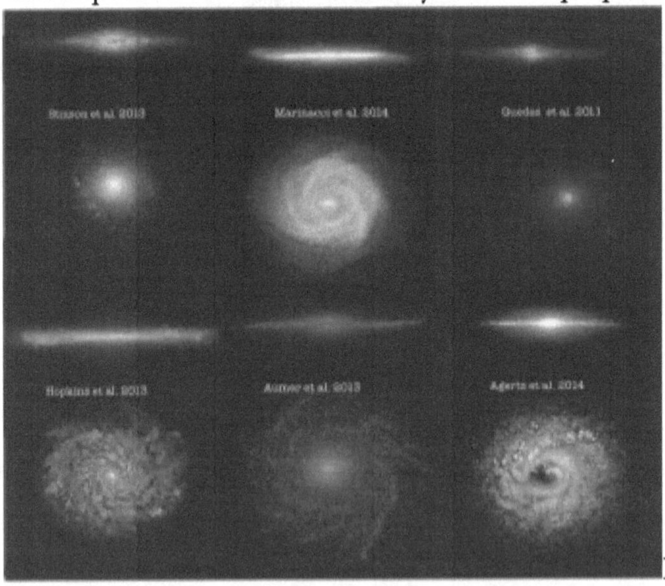

In its simplest form, it states that we observe a universe compatible with our existence because only such a universe can be observed by beings like us. This is logically sound—but many find it unsatisfying. It explains *why we shouldn't be surprised*, not *why the conditions exist*. It addresses observation, not origin.

To ease that discomfort, some physicists propose a multiverse.

In this picture, our universe is one of many, each with different physical constants. Most are lifeless. A few allow complexity. We find ourselves in one of the rare habitable universes simply because only those can produce observers. Fine tuning becomes a selection effect rather than a mystery.

The idea has mathematical motivation in some theories—but remains speculative.

Others argue the constants may be necessary rather than contingent. Perhaps there exists a deeper theory from which these values follow inevitably, leaving no room for variation. What looks finely tuned would then be unavoidable. This approach preserves unity and avoids multiplying unseen universes, but so far, no such theory has been confirmed.

Beneath these explanations lies a deeper tension.

If the constants are contingent, the universe appears improbably fortunate. If they are necessary, then necessity itself becomes mysterious.

Either way, the question shifts rather than disappears. The universe seems structured in a way that welcomes complexity, and no explanation yet removes that sense entirely.

This is where discussions often become polarized.

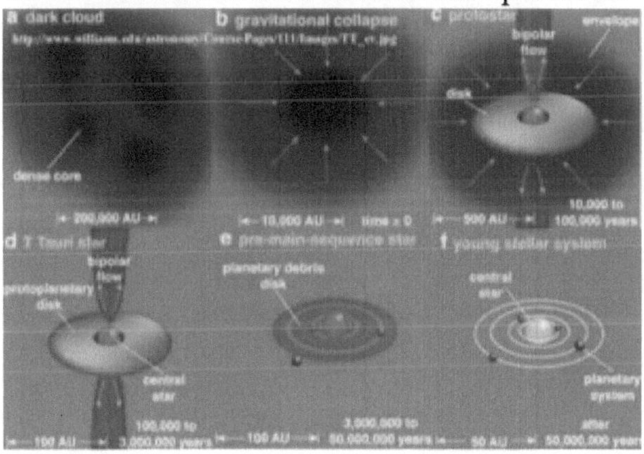

Some interpret fine tuning as evidence of purpose or intention. Others see it as a misunderstanding born of limited perspective. Science itself takes no side. It describes the balance. It does not assign meaning to it. Yet meaning is hard to avoid when the facts themselves feel so suggestive.

It is important to be clear about what fine tuning does *not* claim.

It does not say the universe was designed for humans. Life could take many forms. It does not imply that science has reached its limits. On the contrary, it highlights how much remains unknown. Fine tuning is not a conclusion—it is a question posed by the data.

There is also a human element worth acknowledging.

We are pattern-seeking creatures. When we encounter extreme precision, we instinctively search for explanation. That impulse has driven discovery throughout history. Suppressing it would not make us more scientific. It would make us less honest.

Whether the answer lies in deeper laws, multiple universes, or something beyond current frameworks, the fine tuning of reality stands as one of the most profound facts modern sciences has uncovered. It binds physics to philosophy not by choice, but by necessity.

The universe did not have to be this way.

And yet, it is.

EPILOGUE: The Unfinished Symphony

When a symphony ends, the silence that follows is not empty. It remembers what came before. Themes linger. Meaning doesn't vanish when the final note fades—it changes form. Our journey through the cosmos ends the same way: not with closure, but with resonance. We have traced a story that began before time as we know it, unfolded through fire and form, gave rise to stars, life, and thought, and now pauses—not at an answer, but at a question.

What we have learned is extraordinary. And incomplete.

We have learned that the universe has a history, that it is not eternal in its present form, and that it obeys laws precise enough to be written down and strange enough to remain unsettling. We have learned that simplicity can blossom into complexity, that matter can awaken into life, and that life can become conscious enough to ask where it came from. We have learned that most of what exists is invisible to us, and that what we *can* see is shaped by what we cannot.

We have also learned where understanding runs out.

Science takes us astonishingly far—but not all the way. It tells us how the universe behaves, not why it exists at all. It explains mechanisms with ruthless clarity, yet remains silent on meaning. Philosophy sharpens questions without closing them. Faith offers orientation without coercion. Each illuminate's part of the terrain while leaving other regions in shadow.

This incompleteness is not a defect.

A finished universe would be closed, predictable, and dead to curiosity.

An unfinished universe invites participation. It allows discovery. It leaves room for humility. Mystery is not a failure of knowledge—it is a sign that reality exceeds our frames.

We live at a rare moment in cosmic history.

The universe is old enough to have produced minds, but not so old that everything meaningful has already faded. We inherit the light of ancient stars and the responsibility of fragile futures. Our instruments can see farther than ever before, yet our wisdom has not always kept pace with our power. Knowledge without care is dangerous. Wonder without responsibility is shallow.

The act of wondering itself matters.

To wonder is to pause before reality without trying to dominate it. It is to recognize that existence did not have to include us—yet does. Wonder softens certainty without dissolving conviction. It opens space for learning, gratitude, and care. In a universe that does not demand belief, wonder becomes a freely chosen response.

This book has not tried to settle every tension.

It has not attempted to prove or disprove ultimate answers. It has not crowned science as the final authority or stripped it of meaning. It has not promised certainty where none is available. Instead, it has tried to be faithful to what we know, honest about what we do not, and attentive to what it feels like to be human inside such an immense and intricate reality.

The cosmos does not ask us to be central.

It asks us to be awake.

Awake to the fragility of life.
Awake to the rarity of awareness.
Awake to the weight of choice.

Meaning is not handed to us complete—but neither is it arbitrary. How we treat one another matters more, not less, in a universe that offers no guarantees beyond the present moment.

One day, the universe may grow cold and dark.

Stars will fade. Galaxies will thin. Silence may spread where light once traveled freely. But none of that reaches backward to erase what has already happened. The universe has already known beauty, thought, care, and reflection—because it has known us.

And for now, that is enough.

The story remains unfinished because we are still here.

As long as there are minds that ask, hearts that care, and hands that act with intention, the universe continues to explore itself through us. The final meaning of existence may remain beyond reach—but the act of seeking it is itself one of the most remarkable things the cosmos has ever done.

So, we end not with an answer, but with an invitation.

To keep looking up.
To keep asking.
To live as though it matters.

Because in this vast, ancient, unfinished symphony, you are not an echo.

You are one of the voices still singing.

APPENDIX A: A Walk Through Cosmic Time

To understand the universe is not only to grasp its structure, but to locate ourselves within its immense history. Time is the quiet framework that holds every event described in this book, yet its scale resists intuition. This appendix offers a continuous walk-through cosmic time—not as a table of dates, but as a living sequence of transformations that carried the universe from its earliest moments to now.

The story begins at the edge of what our understanding can reach.

In the earliest fraction of a second, the universe existed in an extreme state of density and temperature. Space and time were inseparable, and the forces that now govern reality had not yet taken distinct form. Our current theories cannot fully describe this era, but it marks the beginning of everything that followed. Expansion began, and with it the possibility of structure, difference, and history.

As the universe expanded and cooled, the fundamental forces separated into the interactions we recognize today. Elementary particles emerged. Matter and antimatter annihilated one another almost completely, leaving behind a slight excess of matter—an imbalance that would eventually become every star, planet, and living body. Within the first few minutes, protons and neutrons combined to form the lightest atomic nuclei, mainly hydrogen and helium. The chemical foundation of the cosmos was set quickly and irrevocably.

For hundreds of thousands of years, the universe remained too hot for atoms to exist.

Electrons could not bind to nuclei, and light scattered endlessly through a glowing plasma. Only when temperatures dropped enough for neutral atoms to form did light finally break free. That ancient radiation still fills the universe today, carrying a faint but precise image of the cosmos as it was in its youth—a fossil record written in light.

What followed was a long cosmic night.

The universe continued to expand and cool, filled with vast clouds of hydrogen and helium. No stars yet burned. No galaxies yet shone. Gravity worked slowly, amplifying tiny irregularities inherited from the early universe. Matter gathered patiently, preparing the conditions for ignition.

After several hundred million years, the first stars formed.

Their light ended the long darkness. These early stars were massive and short-lived, burning intensely and dying violently. In their cores, heavier elements were forged for the first time, then scattered into space through explosive deaths. Galaxies began to assemble along immense filaments of matter, forming the large-scale structure of the universe. The cosmos became luminous and chemically richer.

Over billions of years, stars lived and died in countless generations.

Galaxies collided, merged, and evolved. Heavy elements accumulated.

Around one ordinary star in an unremarkable spiral galaxy, a rocky planet formed at just the right distance to allow liquid water and complex chemistry. On that planet, Earth, life began—not as intention, but as process.

For most of Earth's history, life remained microscopic.

Oxygen slowly transformed the atmosphere. Cells learned cooperation.

Multicellular organisms emerged. Ecosystems formed, collapsed, and reformed. Mass extinctions reshaped evolution again and again. Through it all, Earth remained a small participant in a far larger cosmic unfolding.

Only very recently—within the final moments of a cosmic calendar—did a species arise capable of reflection.

Humans learned to track the stars, measure time, and wonder about origins. In an even smaller slice of time, they developed science, uncovering the deep history that made their own existence possible. The universe began, in a literal sense, to understand itself.

Today, the universe continues its expansion.

Stars still form. Others fade. Life persists on at least one world, carrying memory, curiosity, and responsibility. The present moment hangs between a deep past and an immense future, shaped by both yet owned by neither.

And the story does not end here.

The future stretches into eras where galaxies drift beyond one another's reach, stars exhaust their fuel, and complexity becomes harder to sustain. What remains uncertain is not whether the universe will change, but how meaning—once arisen—will be carried forward within it.

To know this history is not to become insignificant.

It is to recognize participation. You exist at a point where the universe has become aware of its own story. That awareness is brief in cosmic terms, but it is real. And within that brief window, understanding is possible.

The universe has taken a long time to reach you.

APPENDIX B: The Equations Behind the Story

Equations often look forbidding—dense with symbols, unfamiliar notation, and compacted meaning—but at their core they are simply patterns written precisely. Every major equation in cosmology exists because the universe behaves reliably in a particular way, and the equation is our most efficient attempt to say *what that way is*. This appendix is not about learning mathematics. It is about understanding what the mathematics is pointing to.

One of the deepest shifts in modern physics was the realization that space and time are not passive stages on which events unfold.

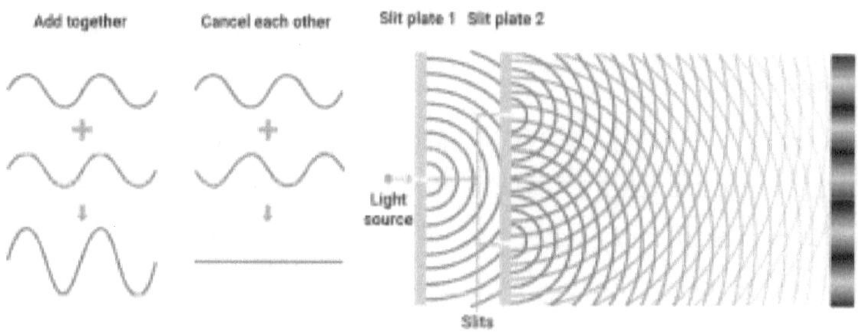

This insight is captured by **General Relativity**, which describes gravity not as a force pulling objects through space, but as the curvature of spacetime itself. Mass and energy tell spacetime how to bend, and that curvature tells matter how to move. Planets orbit, light curves, and clocks tick differently not because of invisible tethers, but because the geometry of reality has changed. The equation formalizing this idea replaces an old mystery—action at a distance—with a new clarity: gravity is shape.

Another central relationship describes the expansion of the universe.

Observations show that distant galaxies recede from us at speeds proportional to their distance, a pattern summarized by **Hubble's Law**.

This is not because galaxies are flying away from a central point, but because space itself is stretching everywhere at once. The equation connects distance and recession speed through a single parameter that reflects the universe's current expansion rate. What matters most is not the number, but the implication: the universe is dynamic, evolving, and has a measurable history written into its motion.

At much smaller scales, equations governing quantum behavior reveal a reality that defies everyday intuition.

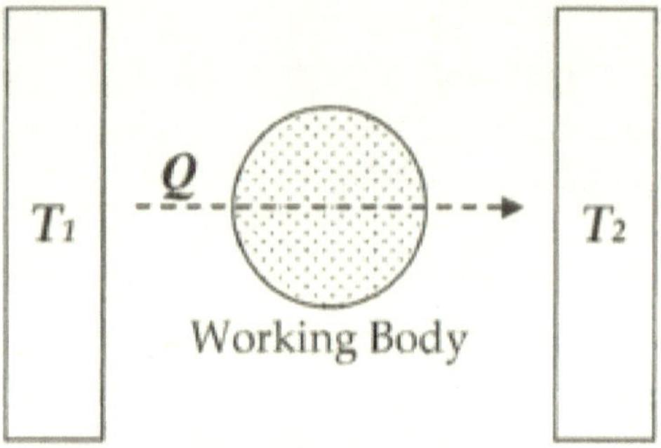

Particles are not described by precise paths, but by probability waves—mathematical objects that encode where something *might* be found if measured. This behavior is captured by equations such as the **Schrödinger equation**, which does not predict single outcomes, but distributions of possibility. This is not a flaw in knowledge. It is a statement about how nature operates at its foundation: uncertain in outcome, exact in probability.

Energy provides a bridge across scales, and one equation in particular reveals how deeply unified reality is.

The relationship expressed in **Mass–energy equivalence** shows that mass and energy are not separate substances, but interchangeable forms of the same thing. Mass is energy held in place. This insight explains why stars can shine for billions of years, why nuclear reactions are so powerful, and why the universe's total energy content influences its long-term fate. It is not about explosions—it is about storage.

Time enters the equations through entropy.

The laws describing entropy do not say that everything must become chaotic in a casual sense. They say that energy tends to spread out, becoming less available to do work. This tendency gives time its direction. It explains why we remember the past but not the future, why causes precede effects, and why the universe changes rather than remaining frozen. Without this principle, there would be no history—and no experience.

What unites these equations is restraint.

They do not explain everything. They do not tell us why laws exist, why values matter, or why consciousness feels the way it does. They describe relationships, not meanings. Yet that restraint is precisely what gives them power. By refusing to say more than can be tested, equations earn the right to say what they do.

It is tempting to think of equations as cold or impersonal, but this misses their origin.

Each equation represents generations of curiosity, disagreement, failure, and refinement. They are cultural achievements as much as scientific ones—products of human attention applied patiently to the world. They exist because people noticed patterns and cared enough to describe them precisely.

Understanding the universe does not require memorizing symbols.

It requires recognizing what those symbols point toward: a reality that is orderly without being rigid, lawful without being lifeless, and open enough to allow novelty, complexity, and reflection to arise. Equations are not the meaning of the universe—but they are among the clearest windows we have into its structure.

The story told in this book rests quietly on these mathematical foundations, even when the symbols themselves remain unseen. They are the grammar beneath the narrative, shaping what can happen without dictating what must matter.

APPENDIX C: Words That Carry the Weight of the Cosmos

Language is the bridge between experience and understanding. Throughout this book, certain words appear again and again, carrying meanings shaped by science, philosophy, and long human reflection. This appendix does not aim to offer dictionary definitions. Instead, it clarifies how these terms are used here, so they remain tools for thought rather than obstacles to it.

Cosmos refers to the universe understood not only as everything that exists, but as an ordered whole. The word carries an implication of structure and intelligibility. When used here, it emphasizes that reality is not a random collection of events, but a coherent system governed by patterns and laws.

Universe denotes the totality of physical existence, including space, time, matter, energy, and the laws that govern them. In many contexts, it refers specifically to the observable universe—the region from which light has had time to reach us since cosmic expansion began.

Observable universe describes the portion of the universe that can, in principle, be observed from Earth because light or other signals have reached us. It does not imply that anything beyond this boundary does not exist. It marks a limit of observation, not of reality.

Big Bang refers to the early phase of cosmic expansion when the universe was extremely hot and dense. It was not an explosion in space, but an expansion of space itself. The term names a boundary in our models, not a complete explanation of ultimate origin.

Spacetime combines space and time into a single framework. Events are located not only by where they occur, but by when. In modern physics, spacetime is dynamic and can be curved by mass and energy.

Gravity is the interaction that shapes large-scale structure in the universe. Rather than acting as a simple pulling force, it is understood as the curvature of spacetime caused by mass and energy, guiding how matter and light move.

Quantum mechanics is the branch of physics describing behavior at the smallest scales. It replaces certainty with probability and reveals a universe that is constrained yet fundamentally open at its foundations.

Entropy measures how energy spreads out and becomes less available for work. It gives time a direction and explains why processes unfold rather than reverse. Entropy does not mean disorder in a casual sense; it describes irreversibility.

Dark matter refers to unseen matter that interacts primarily through gravity. It does not emit or absorb light, yet it plays a central role in the formation of galaxies and large-scale cosmic structure.

Dark energy is the name given to the unknown influence driving the accelerated expansion of the universe. It appears to be smooth, pervasive, and dominant in the universe's total energy budget.

Emergence describes how complex behavior arises from simpler components without external guidance. Emergent properties depend on underlying systems but cannot be reduced to them without losing essential meaning.

Life is used here not as a rigid category, but as a continuum of self-sustaining chemical systems capable of replication, variation, and adaptation. Its origin marks a transition from passive chemistry to active persistence.

Consciousness refers to subjective experience—the capacity to feel, perceive, and reflect. While dependent on physical processes, it resists full reduction to them.

Meaning denotes significance grounded in experience, relationship, and consequence. It does not require permanence or cosmic endorsement, but it does require conscious beings capable of care and responsibility.

Purpose refers to orientation toward an end. In human contexts, it involves intention and value. In cosmic contexts, it remains a matter of philosophical and theological interpretation rather than scientific description.

Faith is understood here not as blind belief, but as trust shaped by reason, experience, and commitment. It operates alongside doubt rather than denying it.

Science is a disciplined method of inquiry grounded in observation, testing, and revision. It explains how processes occur, but it remains silent on questions of ultimate meaning.

These terms form a shared vocabulary for exploring questions that cross boundaries between disciplines. Clarifying them does not eliminate disagreement, but it allows disagreement to be thoughtful rather than confused.

Words matter because they shape how questions are asked.

With a clearer language in place, the final appendix turns outward offering paths for further exploration, reflection, and dialogue recognizing that no single book can contain a universe that continues to unfold.

CLOSING NOTE: To the Reader Who Looked Up

Every book ends in silence, but the silence it leaves behind is not the same for every reader. Some close the final page with answers neatly arranged. Others close it with questions that feel heavier, sharper, and more personal than before. If this book has done its work, you are likely carrying the second kind.

That is not a failure of explanation.

It is a sign of engagement.

The universe does not yield itself all at once. It reveals patterns slowly, often at the cost of certainty. To understand it requires patience, humility, and the willingness to remain unfinished. The same is true of understanding ourselves. We are shaped not only by what we know, but by what we are still learning how to ask.

You did not read this book from outside the cosmos.

You read it as something the cosmos produced. Your curiosity, your doubt, your wonder—even your frustration—are not add-ons layered onto reality. They are expressions of a universe that has grown complex enough to question itself. In you, atoms that once drifted through interstellar space have learned to wonder why they exist at all.

That fact deserves respect.

If you are a scientist, may this journey remind you that precision and humility are allies, not opposites. If you are a person of faith, may it remind you that wonder is not a threat to belief, but one of its deepest sources. If you are neither, may it still affirm that meaning does not require permission to exist.

No worldview emerges unchanged from an honest encounter with the cosmos.

Certainties soften. Responsibilities sharpen. What remains is not despair, but seriousness—a seriousness that recognizes how rare awareness is, how fragile life can be, and how much depends on what conscious beings choose to do with the time they are given.

The universe does not promise us significance.

But it has given us the capacity to create it.

If, after reading, you find yourself looking at the night sky a little longer than before, if familiar arguments feel less satisfying than genuine questions, if care feels harder to dismiss—then the journey has not ended. It has begun again, this time not on the page, but in the way you stand within the world.

So, take this book as it was intended.

Not as an answer.

But as a companion for the walk.

And when you look up, remember this:

The universe is vast.

Your life is brief.

And what you choose to do with your awareness still matters.

ACKNOWLEDGMENTS: Gratitude in a Vast Context

No work like this is ever created alone, even when a single name appears on the cover. Every idea explored in these pages' rests on generations of thinkers who asked difficult questions, tolerated uncertainty, and refused to stop looking simply because the answers were incomplete. Scientists, philosophers, theologians, poets, and countless ordinary people who paused to look at the night sky all contributed—knowingly or not—to the inheritance this book draws from.

Gratitude is owed first to those who expanded humanity's understanding of the universe through disciplined curiosity. To astronomers who measured faint light across impossible distances, physicists who trusted equations before experiments could confirm them, biologists who traced life's deep roots, and thinkers who insisted that clarity and humility must grow together. Their work reminds us that knowledge advances not through certainty, but through courage guided by restraint.

There is also gratitude for traditions of thought that preserved wonder alongside reason. To religious scholars who treated observation as devotion rather than threat. To philosophers who refused to reduce meaning to mechanism alone. To educators who taught that asking honest questions is not an act of rebellion, but an expression of respect. Without these traditions, the dialogue explored in this book would collapse into either dogma or despair.

On a more personal level, gratitude belongs to those who make reflection possible. To teachers who encouraged questions rather than memorization. To readers who value depth over speed. To conversations that challenged assumptions without dismissing sincerity. To silence itself, which allows ideas to settle, connect, and mature.

Finally, gratitude is owed to you, the reader.

To read slowly in an age of haste is an act of resistance. To sit with uncertainty without rushing toward closure is an act of courage. To take the universe seriously—not as spectacle, but as context—is an act of responsibility. By engaging with these pages, you participated in the very process the book describes: the universe becoming aware of itself through thoughtful attention.

If this work contributes even a small measure of clarity, humility, or wonder, it belongs as much to those who received it as to the one who wrote it.

In a cosmos shaped by interconnection, gratitude is not merely a courtesy.

It is an accurate description of how anything meaningful ever comes to be.

www.ingramcontent.com/pod-product-compliance
Lightning Source LLC
LaVergne TN
LVHW090429210926
848480LV00002B/757